A ship is a big boat that can carry passengers or cargo over the sea.

The front of a ship is called the bow. The back is the stern.

The ship's body is called the hull. The keel is the backbone of the ship. It runs along the bottom.

A long time ago, ships used the wind to move across the sea. Sailing ships had sails made out of cloth and a tall mast.

A container ship carries cargo. Cargo can be things like coal, grains, farm products.

Passenger ships can be called cruise ships. Cruise ships travel around the world.

An icebreaker ship breaks through the ice. Ships can be all kinds of sizes and shapes.